SCIENCE MUSEUM

The Challenge of the Chip

W H MAYALL

LONDON: HER MAJESTY'S STATIONERY OFFICE

The beginning

The word **chip** means in general a small piece of something solid. In recent times it has come to mean in particular a tiny piece of the element silicon, which, because of the way we have learned to pack thousands of electronic functions into it, is changing the way we live and work and challenging us to ensure that the change is for the better. In a chip five millimetres square and half a millimetre thick can be put the main parts of a computer that fifteen years ago would have filled a room five metres square. We are dealing with micro-electronics. The history of the discovery of the electron and of the study of the way it moves in crystals goes too far back to tell in this short book. The part of the story that matters here begins around 1947 in a research laboratory and develops with the growth of space flight and of advanced defence systems.

The history of computing goes back to the mid-nineteenth century when Charles Babbage designed (but never completed) his 'analytical engine'. This was to be more than just a machine for rapidly doing tedious and complicated arithmetic. It was intended to **store** the results of one calculation for use in another: it was to have a **memory,** which is what makes possible a problem solving computer.

In 1904 Ambrose Fleming invented the diode, in which electrons emitted from a heated wire in an evacuated enclosure were attracted to a positively charged plate. The current could travel in only one direction so the diode could be used as a non-return valve to rectify an alternating voltage. In 1904 Lee de Forest devised the triode which **controlled** the flow of current. This became the basis of all future develop-

2 Charles Babbage (1791–1871), a Cambridge mathematician, whose Analytical Engine might have been the first computer.

ments in radio, and in other electronic processes. One
important discovery was that a combination of triodes
could act as a switch, a device that could, in effect,
say 'this way/that way', which is equivalent to counting
up to 2.

In 1945 ENIAC (Electronic Numerical Integrator and
Calculator) was built at the University of Pennsylvania.
It was significant for three reasons. It was the first all-
electronic calculator (but not a computer since it had
no memory). It used valves for operating switching
systems to undertake its calculations – no less than 18,000
of them in a system which weighed 30 tonnes overall.
And it could not have used any more valves because,
owing to the unreliability of valves, engineers might
have had to spend twenty-four hours a day replacing
those which failed!

Nevertheless valves for switching purposes were
employed by scientists and engineers to create the first
all-electronic computers. These began to appear in the
late nineteen-forties and were available commercially
in the early nineteen-fifties. They fascinated and
frightened the public. There were many debates about
whether they were 'intelligent' largely because they
could be used to plan the best way in which to undertake
complex tasks or to provide their users with large
amounts of information as well as to undertake complicated calculations. Most debates ended, however,
with the assertion that one only got out what one put in :
a computer could do no more than it was told to do.
'Rubbish in – rubbish out' was the usual conclusion to
any such argument. It still is.

3 A radio valve of 1910, forerunner of the
valves used in the first electronic computers.

4 The experimental computer built at
Manchester University upon which the Ferranti
Mark I Star was based. See cover caption.

5 John Bardeen, Walter Brattain and William Shockley (seated) – inventors of the transistor.

6 Facsimile of the first transistor invented in 1947.

7 The first integrated circuit invented in 1958 by Jack Kilby.

The transistor

In 1947 three scientists working at the Bell Telephone Company's Laboratories in the United States invented a totally novel device which became known as a **transistor.** It was not an accidental invention but the culmination of a long line of study of the electrical properties of crystals going back to the mid-nineteenth century.

Nevertheless, as compared with previous practical achievements in this field, the creation of the transistor by John Bardeen, Walter Brattain and William Shockley was a very considerable step forward. The transistor could do what Lee de Forest's triode valve had done but it did not need any part to be heated to make it work. It was solid. It was much smaller.

Bardeen and Brattain invented the point contact transistor which had two wires carefully positioned on a crystal of germanium. Shockley followed up shortly afterwards by creating the bipolar or junction transistor in which there were no wires. In doing so, he initiated microelectronics.

Because of their comparatively small size, the first transistors were used to replace valves in hearing aids but, by the middle of the nineteen-fifties, the first transistor radios had appeared. It took a little longer for transistors to appear in computers. Computers use thousands of switches. Various improvements to transistors were made, one being the replacement of germanium by silicon. Transistors, with their small size and robust qualities, would evidently suit computers far better than thermionic valves. It was the merits of computers themselves which had still to be appreciated.

Attitudes changed rapidly after the mid-nineteen-fifties and computers with transistors had appeared in industry by the end of the decade. In the mid-fifties also they began to be employed in the complex control systems used in defence equipment and rockets. For such applications smallness and utter reliability were vital requirements and, with its commitments to defence and later to entering space, the United States government began to finance research and development work on miniaturising electronic systems.

The limit to miniaturisation was set by the fact that the components of circuits were individual items that had to be wired together. Wiring was bulky. In addition it was unreliable because its quality was dependent upon the operations of human beings with different degrees of skill. The solution lay in integrating all the components in one solid piece of material.

The integrated circuit and the chip

In 1952, G. W. A. Dummer of the Royal Radar Establishment in Great Britain put forward the concept of an integrated circuit, but his ideas were not followed up. The effective integrated circuit was to be an American achievement. Engineers in several companies developed ideas for integrating a number of transistors into a single package and Jack Kilby of Texas Instruments Incorporated made the first working circuit in 1958. But, a new company, Fairchild Semiconductors, set out in 1957 to develop a novel concept in miniaturisation; namely the planar process in which transistors were created in the surface of a wafer of silicon. Success came by 1960. The planar process set scientists and engineers on a hectic race towards previously unimagined degrees of miniaturisation and complexity in microelectronic circuits.

Integrated circuits were first used in defence equipment and space vehicles but, by 1963, they were being used in commercial products, at first in small personal things like hearing aids, but soon afterwards in computers. These, in defence and space equipment then in business and industrial equipment, were to be the greatest users of integrated circuits.

A clear aim was to increase the number of transistors on a chip so that more computing capacity could be built into a given space. Fortunately the achievement of this by the planar process also made for cheapness and, by eliminating wiring, reliability.

In 1965, 30 components could be put on a 5 millimetre square silicon chip: by 1975, 30,000; by 1978, 135,000. But numbers have not been the only gain; the minute circuits are far more complex. The computer which filled a large room thirty years ago is now on a 5 millimetre chip.

8 Part of an ICT 1900 series computer built in the mid-1960s using integrated circuits.

9 A modern microelectronic circuit, shown in the **smallest** square, can have more than 250,000 components.

10 Electronics is commonplace in our daily lives. Microelectronics soon will be.

11 Prestel, the Post Office's information service by television, uses microelectronic systems.

Electronics everywhere

How much has electronics and microelectronics brought within our reach? Indeed how much microelectronics do we already take for granted?

The telephone, invented by Bell in 1875, is now vital to the ways in which we work and live. Sound broadcasting, forecast by Hertz's demonstration of electromagnetic waves in 1887, became available to millions of people in the nineteen-twenties and thirties largely as a result of the invention of the thermionic valve at the turn of the century. Indeed it remained the key component in radio until transistor sets appeared in the mid nineteen-fifties.

Television, with its beginnings in the nineteen-thirties, became a wide inescapable influence from the nineteen-fifties onwards. In the same period electronics spread with increasing speed to control systems and computers. When automation, which means the automatic control of production processes, began to be employed, it was electronics which increasingly provided the instruments to measure the performance of these processes so that they could be controlled. Electronics also helped to produce the control mechanisms themselves.

It was not long before the design of aircraft and rockets exploited all three applications of electronics, namely communication, control and computing – resulting in vastly improved navigation and control systems.

Seen initially as a device for solving complicated mathematical equations more quickly than we could solve them ourselves, the computer had become as indispensable as the telephone in most large industrial and commercial organisations by the end of the nineteen-sixties. In addition, it had come to be accepted, and indeed relied upon, in many areas of public administration including tax collection, pensions and population records.

Microelectronics in many places

Computers may still be thought of as large and daunting devices even by a person who has just bought a programmable pocket calculator containing all the essential features of a computer. Now they are everywhere.

There has been a curiously prestigious character about some electronic products. The nineteen-thirties radiogram (radio receiver and record-player united in an imposing cabinet) and more recently the remote controlled television set have been status symbols for a time, but have soon become widespread. Digital watches were status symbols when they first appeared in the late nineteen-sixties, but the watch with half-a-dozen functions is now everywhere. So is the pocket calculator, which in less than a decade has become cheap enough to be used by everyone. Games and puzzles employing calculator principles are now big business. Some have real educational value, as have some of the many games that one can play on the television screen.

Information services linked to television are being developed including **Prestel** operated by the Post Office, **Ceefax** by the BBC, and **Oracle** by ITV. They present a wide variety of information which the user can summon on to his television screen, such as weather forecasts, share prices and sports results as well as world news.

Following personal calculators will soon come personal computers with systems for household management and accounts, for self education and for storing personal and private information.

How much more can we expect to see and be asked to accept in the next decade and beyond?

12 Pocket calculators which depend upon microelectronic chips are now commonplace. This one also acts as an alarm clock **and** a musical instrument!

13 A large number of microelectronic toys and games are now available. This chess set speaks the opposing moves.

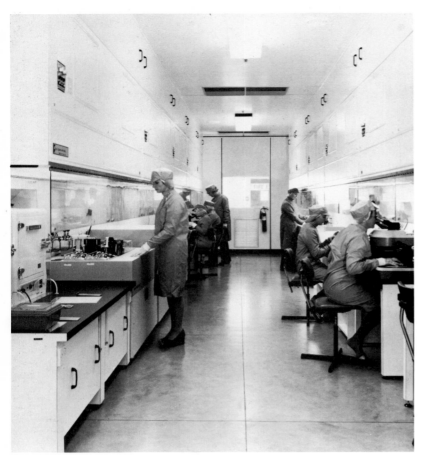

14 Typical working conditions for making chips at Mullard's factory in Southampton.

A new industry

The transistor was invented, developed and first manufactured on a large scale in the United States. Existing valve manufacturers played little part in this development, seeing in the transistor a threat rather than an opportunity. Owing to various limitations the early transistor could not act as a direct replacement for the familiar valve, so the Bell company in whose laboratories it had been invented set out to improve it and to grant licences to other organisations. Raytheon, with its transistorised hearing aids, was the first company to exploit the transistor commercially.

About 15 companies set up manufacture of hearing aids in the United States between 1951 and 1956, and about 25 were making transistors by 1956. Between 1954 and 1956 the output of transistors was about 28 million, but this number was small compared with the 1,300 million thermionic valves manufactured in the same period.

This huge production was due to an enormous increase in demand for all sorts of electronic equipment. Many scientists and engineers inside the big companies foresaw the future of the transistor better than did their own managements and undoubtedly there was reluctance to adopt transistor technology on the part of those technical managers who had become committed to the rigid systems of their companies and to customary methods of manufacture.

Thus a number of substantial valve makers began to lose their experts who either set up their own companies or joined small ones that were more alive to the opportunities which the transistor presented. Texas Instruments was a tiny company in 1951 with no great experience in electronics. But it set itself three objectives and took on staff, notably from Raytheon, to achieve them. The objectives were to make a silicon transistor, to produce the first transistor radio and to devise ways of making silicon so that every bit of impurity was removed. The latter objective was, and still is, crucial to the manufacture of silicon transistors and, of course, of integrated circuits.

New concepts create new companies

Small companies blossomed and took staff from the conservative valve makers. Raytheon lost people to Transitron, a company set up in 1952, which became the second largest maker of transistors within three years. Bell Laboratories lost the inventor of the bipolar transistor, William Shockley, who set up his own company in 1955 in the Santa Clara valley near San Francisco.

These companies in their turn lost scientists and engineers who set up their own companies. The Santa Clara valley became the home for so many companies working on transistors and integrated circuits that it came to be known as Silicon Valley. The most significant was Fairchild Semiconductor. The eight scientists who started it in 1957 had all come from Shockley's organisation, to explore the technologies needed to make integrated circuits. By developing the planar process in 1960 they established the future pattern of production.

Larger companies like Bell also contributed to developments. In addition there was, and still is, much open exchange of information between the companies themselves, and between these companies and the great American technological universities.

But it was in Silicon Valley that a vital new industry was created. Scientists and engineers moved from company to company or set up their own. The name 'Fairchildren' was ascribed to those who left Fairchild Semiconductor to set up or join new companies, which, in due course operated in the same area. Few of the traditional valve companies could keep up with these hectic developments and even the first transistor makers found it difficult.

The world now follows where America has led. Japan, in particular, under forceful government direction, has become a major producer of microelectronic systems. In Britain, efforts are being made to establish a microelectronics industry with government support. The British companies, Ferranti, Mullard, Plessey and GEC manufacture integrated circuits. Indeed Ferranti produced the first microprocessor in Europe.

No industrial country can now manage without a microelectronics industry unless it decides to depend upon another country. An entirely new science-based industry has appeared in less than twenty years.

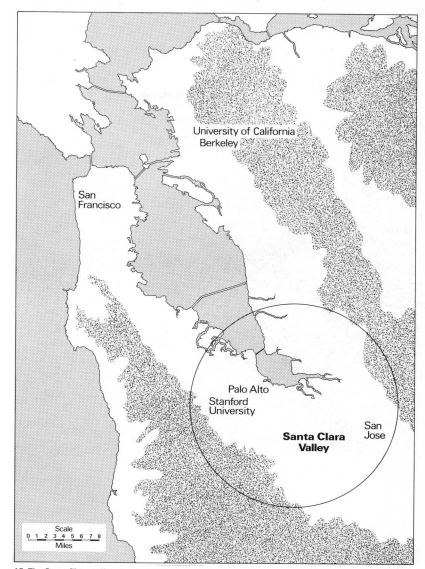

15 The Santa Clara valley, birthplace of the microelectronics industry, which became known as 'Silicon Valley'

9

Threats and promises

16 The dole queue symbolised large-scale unemployment in the early 1930s. Will microelectronics help to bring it back?

17 Automatic systems and robots like this one which is feeding three machine tools will take over routine jobs.

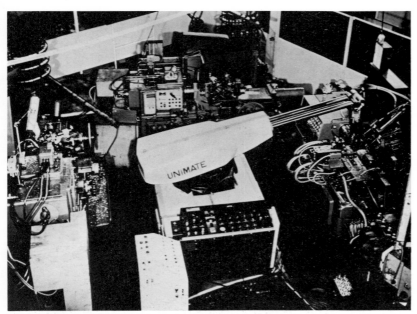

History can produce many instances of people fearing the effect on their livelihood of new inventions which could do their work faster or more efficiently. The best known British example is provided by the Luddites, workers who rioted to destroy textile machinery between 1811 and 1816. At the present time there is a great deal of discussion about the danger to employment as well as about the promise of increased prosperity, better working conditions, and more leisure that microelectronics can help to create. Here are some areas which need thought.

In a country which depends upon selling manufactured goods a technical development such as microelectronics which cuts manufacturing costs and increases sales has to be taken seriously. Microelectronics may also improve quality by ensuring consistency. In addition it makes possible the control of long sequences of operations on products, imitating human actions enough to justify the use of the word **robot** for systems now being used in making cars, television sets or washing machines. These are made in long runs, but microelectronic control can also apply to small batches, and even with suitable programming, to one-off special orders. This means that labour can be released for tasks better done by a person than a machine.

So much for the factory floor, which has been affected by modern automation since the nineteen-fifties. But the office worker is also affected for there are small computers which can provide swift communication, assemble and print standard letters automatically, and handle accounts in new and more efficient ways. Managers may be affected where large amounts of information can be assessed better by computers than by people. Even designers are finding some of their functions better executed by computers.

So far as can be seen, the only threat of microelectronics is to employment. It does not of itself threaten us with pollution, noise or ill-health. Indeed it can be applied to relieve the dangers of other aspects of technology. As for unemployment it seems that microelectronics is likely to create new opportunities for work. Scientists, engineers and technicians will be needed to design and manufacture the new devices, and people of many levels of skill and training to use and apply new machinery in new work situations. The hope is that it is not rewarding work that will be reduced, but drudgery and unnecessary hard, undignified or disabling labour.

A well-informed society?

We spend less time at work than we did thirty years ago. In thirty years time we shall spend even less. At present, leisure is for many a respite from something they have to do in order to exist; it is a time for rest and for casual amusements and passing sensations. Pessimists argue that when work takes up less time and becomes less demanding we shall become aimless, incapable of expressing ourselves, and dulled by trivial entertainment coming to us mainly through our television sets! Optimists say that the division between work and leisure will disappear. We shall have more chance to exercise our creative and cultural talents, more time to assist in the improvement of society and more incentive to learn about ourselves and the world about us.

It is in providing the means to learn that microelectronics may best serve us. Prestel and personal computers are but the forerunners of systems that will enable us to learn when we want to rather than during prescribed periods, and will provide us with many more sources of information. Microelectronics will create what is coming to be known as the 'information society', a society in which we shall be able to communicate knowledge and experience far more easily than we can now. Five hundred years ago Renaissance culture was disseminated by the newly born printing press. Microelectronics stands ready to spread the culture of today and tomorrow.

18 More production systems will be controlled by computers with operators monitoring performance. MINOS, shown here and employed by the National Coal Board, works in this way.

19 We often regard leisure as a relief from tasks we must undertake to maintain ourselves. Given more 'leisure' time, our attitudes may change.

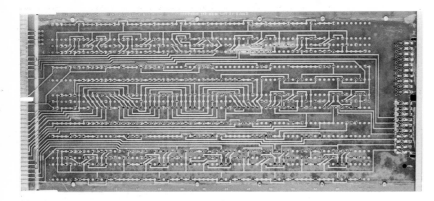

20 A large printed circuit board for telephone exchange equipment built by ITT Europe ; the printed circuit being on one side of the board as shown, the components being mounted on the other side.

21 An integrated circuit may look like a printed circuit board, but all its components are contained in one material–silicon. Those in this simple circuit comprise transistors (T), resistors (R), diodes (D) and capacitators (C).

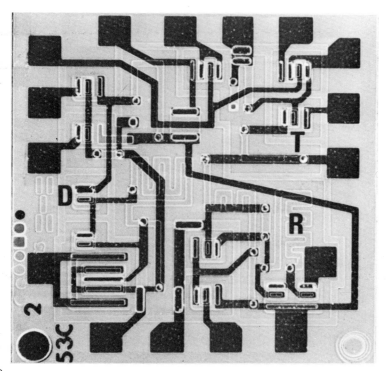

The chip

Through a microscope a chip looks like nothing more nor less than a tiny printed circuit board. Printed circuit boards began to appear in the late nineteen-forties and early fifties in order to overcome the high cost and uncertain reliability of joining electronic components together with wires. Components were mounted on insulating boards which had been literally printed with the circuits in which the components operated.

Today, almost all electronic products made in any quantity contain printed circuit boards. The components are usually mounted on one side of a board and the circuit is printed on the other side. The transistor being smaller and more robust than the thermionic valve suited this type of construction. Solid and squat it can be found as an independent component on the printed circuit boards of our radios and television sets.

There is, however, one important and enormous difference between the way the transistor, and indeed any other component, is installed in a printed circuit and the way it is contained in an integrated circuit. All the components on a printed circuit board are in individual containers whereas all the components in an integrated circuit are in one container ; namely the material silicon. **Moreover they depend upon this material in order to work.**

Transistors to operate switching systems, resistors to resist the passage of electricity and capacitors to store an electrical charge all use silicon in order to function. And they can all be created at the same time in one chip of silicon which, while being much smaller than any printed circuit board, can also carry very many more components.

Silicon is not the only material from which the chip might be made but it is by far the most important. It is one of a group of materials classed as semiconductors. These are materials which would not normally allow the passage of electricity but which can do so after they have been treated in various ways. For example, silicon will allow the passage of an electrical current after minute amounts of other elements are added to it. This process has come to be called 'doping' and its effect can be understood by seeing how a semiconductor diode works.

Silicon – the semiconductor

A crystal whether of a metal (which conducts electricity well) or of silicon (which does not) is made up of atomic nuclei held together in a fixed lattice by associated electrons. In a metal many electrons can move easily and so it conducts. In pure silicon the electrons are relatively fixed, so it is a poor conductor. But if there are irregularities in the crystal lattice (caused, for example, by the presence of a few atoms of another element) current can pass. If some atoms are replaced by phosphorus the material has electrons surplus to the number needed to maintain the crystal structure. Boron creates a deficiency of electrons so there are what we can think of as 'holes' in the systems where electrons ought to be, (the holes being positive because the electron is negatively charged). Silicon 'doped' (as one says) by adding phosphorus is called **n** type ; and by adding boron it is called **p** type.

By using these two types together one can create all manner of electronic components such as transistors, diodes, resistors and capacitors. The most basic method of use is to put **n** type against **p** type to make the simple diode. Electrons in the **n** type tend to move towards the holes in the **p** type and vice-versa. But this movement is opposed. Some electrons are captured by holes and cancel each other out. Also electrons by quitting the **n** type leave it positively charged **on balance**. Movement of 'holes' leaves the **p** type negatively charged **on balance**. Eventually an equilibrium is reached. Now, if a positive electrical potential is applied to the **p** type side, the equilibrium is disturbed.

Holes are repelled across the junction to the **n** type side, and electrons move towards the positive **p** type side ; a current flows. But if a positive potential is applied to the **n** type side electrons and holes are each held back at their normal positions, and virtually no current flows.

This means that current can flow in only one direction : the **n** type **p** type junction, like Fleming's diode, behaves as a non-return valve. A junction transistor can be a sandwich of **n** type material between two **p** type materials or vice-versa and it can act both as a switch and as an amplifier.

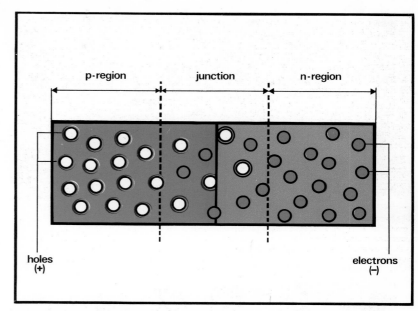

22 Simple diagram of a junction diode showing **n** type and **p** type materials.

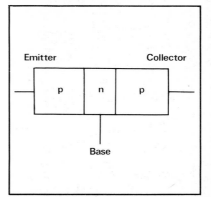

23a Schematic representation of a junction transistor which can be p-n-p as shown, or n-p-n.

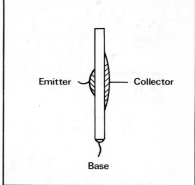

23b Typical construction of an early junction transistor.

24 Planar construction in a silicon chip for different types of component. Code: red for **p** type silicon, light blue for **n** type silicon, dark blue for **n +** type silicon (heavily doped), brown for silicon dioxide and white for metal interconnections.

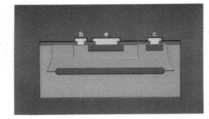

Transistor

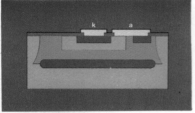

Diode

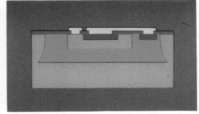

Capacitor

Resistor

MOS transistor

The planar process

The original transistor invented in 1947 was called a point contact transistor. It comprised a piece of semi-conducting material, at that time germanium, to which wires had to be carefully connected. The relative positions of these wires determined the performance of the transistor. Positioning was so critical that no two transistors behaved in the same way. While better manufacturing techniques improved the situation, germanium was never a satisfactory material being adversely affected by changes in temperature.

The junction transistor, which Shockley introduced, did not have the same manufacturing problems. Wires were required only for connecting the transistor to the rest of an electronic circuit. They played no part in the way the transistor worked. This depended entirely upon how **n** type and **p** type materials were placed relative to one another and upon the extent to which they had been doped. Constructed in the form of a sand-wich, the junction transistor could never be other than an individual component. The planar process was to change all this. The **n** type and **p** type materials, instead of being sandwiched together, were created in precisely defined regions adjacent to one another on one face of a piece of semiconducting material. While transistors could be and are produced as individual components by this process, it enabled more than one component to be produced on one piece of material. And these components need not just be transistors.

Resistors, diodes and capacitors can also be pro-duced by the planar process and other types of transistor can be created. Moreover they can all be created on one face of a silicon chip at the same time.

After the planar process came into use, the number of components that could be created on a chip doubled annually. The planar process is a triumph of technology rather than of science. It was encouraged by the need to miniaturise electronic circuitry. But, at the same time production economics played an important part. Clearly the more components that can be produced from a given amount of material without increase of production cost, the cheaper they will be. The planar process satisfied the user of electronic systems and the maker. It ranks as the next important stage in the development of microelectronics after the invention of the junction transistor.

Getting the most out of chips

Putting more capacity into a smaller space is of enormous help to the advanced control systems employed in defence systems, space travel and complex industrial processes. And putting more components onto a chip clearly aids this process. But it does more. We noted initially that wiring is potentially unreliable. Consequently the more wiring we can get rid of, the more likely are we to be able to improve reliability. And while the reliability of equipment is of considerable importance in military systems it is of growing importance to ourselves.

Reliability may also be increased because enormous numbers of electronic components are employed in modern systems and variations of performance between individual components are more probable than between integrated components. So the more functions we can fit into a chip the more reliable will they be. In addition, we shall see in the next section that getting more chips out of a 'wafer' tends to improve reliability still further.

It is clear that integration reduces enormously the cost of making up a circuit owing to the reduction in the amount of assembling and wiring involved. Moreover, due to the methods of manufacture employed, the cost of producing a transistor on a chip is very much less than that of producing an individual transistor. And this, of course, applies to other devices like resistors and capacitors. Finally, because so much more electronic capacity is being built into a much smaller space, the cost of making mounting systems, whether they are printed circuit boards or other arrangements, and the cost of casings and enclosures are also reduced.

Indeed it is the low cost of the chip as much as its capabilities that enables us to employ it in situations undreamt of even ten years ago.

So we have moved from the small scale integration of the early nineteen-sixties, amounting perhaps to 16 components on a chip, through medium scale integration to the large scale integration of today. Very large scale integration is about to begin.

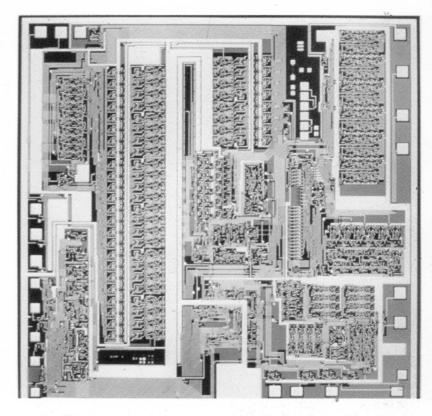

25 A chip embodying 15 different functions and measuring about 6 millimetres square designed for use in a counter.

26 The number of components on a chip has increased rapidly since the first integrated circuits were manufactured.

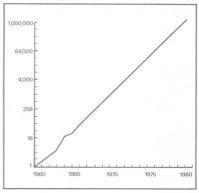

27 Section of a silicon crystal when first produced.

28 After being ground completely smooth, the crystal is cut into wafers which must also be smooth and flawless.

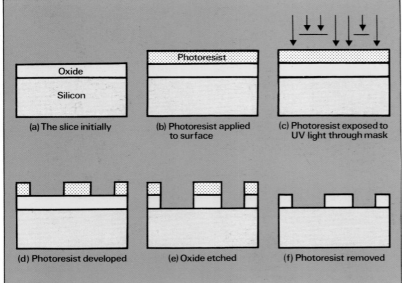

Oxide / Silicon	Photoresist	↓↓↓↓
(a) The slice initially	(b) Photoresist applied to surface	(c) Photoresist exposed to UV light through mask
(d) Photoresist developed	(e) Oxide etched	(f) Photoresist removed

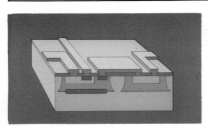

29 Stages leading up to 'doping' the silicon to produce n type and p type regions using masks to create apertures or 'windows' through which the doping occurs.

30 Masking and doping continue until all components have been produced. Aluminium is then used to connect them together into the circuit.

Making chips

Chips are cheap when we consider what they do. But the equipment required to make chips is anything but cheap. Every process must be controlled precisely and carried out in conditions of absolute cleanliness. A speck of dust or a minute amount of impurity in the silicon can stop an entire production sequence.

Pure silicon is essential. It is produced by treating chemically the main constituent of sand, silicon dioxide until it is 99.9999999 per cent pure. Then it is melted under an inert gas and a seed (a small crystal of pure silicon) is used to start a crystallisation process.

A crystal begins to form and is drawn from the melt bit by bit. It is not crystalline in external appearance, being cylindrical, about seventyfive to a hundred milli-metres in diameter, and at least a metre long. Never-theless the whole bar is one single crystal of silicon and ideally its atomic arrangement is perfectly regular throughout. The crystal bar is then ground to an accurate diameter and cut into wafers whose surfaces are also ground and polished to a mirror finish. The wafers can be from a half to a quarter of a millimetre thick and must be completely free of scratches, marks or any other defect. They are the foundations upon which the chips are formed.

Masking and doping

Depending on their size, up to five hundred chips are formed side by side on a wafer. To do this, highly re-fined photographic techniques are employed, together with techniques to introduce into the silicon the ele-ments which produce n or p type regions (doping).

The first stage is to produce a very large drawing of a single chip. From this drawing are worked out a number of 'masks', each one defining the tiny areas where a particular type or intensity of doping is required. These masks are then photographically reduced and repro-duced on a glass photo-mask in the pattern in which the chips will be made on the silicon wafer.

A series of processes then follows which build up the chip structure layer by layer. This may take up to twelve weeks to complete. We shall briefly describe just one stage here, which consists of the photolitho-graphic techniques which define the areas on the chip to be doped, and then the doping itself.

First, a layer of silicon dioxide is carefully formed on one side of a wafer of clean and pure silicon. The surface of the silicon dioxide is then coated with a light-sensitive material called photo-resist. A photo-mask is placed over the top of the photo-resist and the wafer is exposed to ultra-violet light through the mask.

The masked areas are not affected by this light but the photo-resist hardens elsewhere. Thus a pattern of the areas where doping is required is imprinted on the silicon dioxide after treating the wafer in a solution of developer which removes the unhardened photo-resist.

The wafer is next immersed in an acid which also removes the silicon dioxide while leaving the hardened photo-resist and the pure silicon base unaffected. After the remaining photo-resist has been removed, the wafer is ready for doping. This consists of exposing the wafer to an atmosphere containing either phosphorus or boron under conditions of high temperature. Atoms of the dopant then diffuse through the 'windows' in the silicon dioxide layer created by the masking process.

This completes the first stage and the whole process begins again until all the layers are completed. Finally after any remaining photo-resist has been removed, masking techniques are employed to enable a pattern of aluminium to be deposited which will connect the **n** type and **p** type regions into a circuit. Silicon dioxide is also deposited to act as an insulator.

The final chip

Before the wafers are cut up into chips, each is thoroughly tested and those which fail are marked so that they can be discarded. The proportion which fails can be up to 70 per cent, caused both by manufacturing faults and by the remaining basic defects in the silicon structure. The smaller chips are less likely to be affected by the flaws in the silicon so the more chips there are to a wafer, the higher is the proportion of good chips.

Each chip is then housed in a robust plastic package, gold wires being employed to make connections to pins at the base of the package. Wiring is done automatically or semi-automatically under the same conditions of utter cleanliness that have applied to all preceding processes.

After further testing the final packaged chip is ready for assembly on a printed circuit board. It has been built by expensive equipment. But it is cheap because it has been made in quantity – very cheap when one considers that the basis of three hundred computers might easily be created on one wafer alone.

31 Every chip on a wafer is tested and those which fail are marked for rejection.

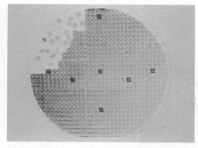

32 The wafer is cut into individual chips.

33 The chip is placed in its package and connected by gold wires to pins in the base of the package. After further testing it is ready for use.

34 A variety of standard packages are shown here.

35 Designing microprocessors cannot now be
undertaken without using computers which
will probably contain microprocessors!

The microprocessor

The microprocessor is so special a type of chip that
many people regard it as **the** chip. It is nothing less than
the heart of a computer – the central processor – on a five
millimetre square piece of silicon.

The first integrated circuits were made for particular
duties, each design of circuit being dedicated to one
distinct task. These 'dedicated circuits' are still produced. However, early in the nineteen-seventies, Intel,
a 'Fairchildren' company, was approached to design
integrated circuits for a family of electronic calculators.
Making a special integrated circuit for each calculator
would have been expensive, so Intel wondered if a
single circuit could be designed to suit all the calculators.
They were thinking, in effect, of producing miniature
computers. And what they developed finally was the
central processor of a computer – a device which could
be programmed to work not only in each calculator **but
in other types of electronic product.** They called it a
microprocessor to identify its role as a miniature central
processor. But what exactly is a central processor?

Components of a computer

We said that a computer has a memory, that is a place where information can be stored and from which it can be withdrawn. The information is used for computing operations by an 'Arithmetic and Logic Unit' (ALU).

The instructions for these operations are fed to the computer by various methods, for example by a type-writer-style keyboard which, instead of typing letters sends electrical impulses into the system. The instructions must be translated since they will be literally in a different 'language' from that employed by the computer to do its work.

A control unit acts on the translated instructions to operate the ALU and to fetch information from the memory when required. It employs a timer or clock to make sure that all operations are conducted in an orderly fashion.

The central processor of a computer must include a control unit and an ALU. But it will usually contain other systems, such as 'buffers', which are temporary memories needed because different parts of a computer operate at different speeds. For example, the ALU can produce answers very much more quickly than these answers can be presented to us as a print-out or on a visual display unit (VDU). So the buffer takes answers and sends them to the typewriter or VDU at the appropriate speed.

Finally, the central processor may house those parts of the computer's memory which are used most frequently ('registers': time-savers for complicated tasks).

A complete computer consists of a central processor and a main memory together with an input and an output device which may be a keyboard and a VDU respectively.

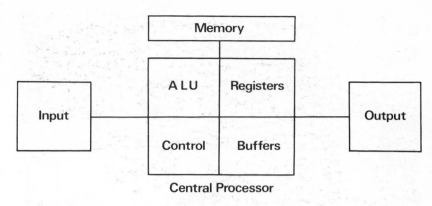

36 The basic features of a computer showing what may be contained in the central processor.

37 A microprocessor made by the Intel Corporation, USA.

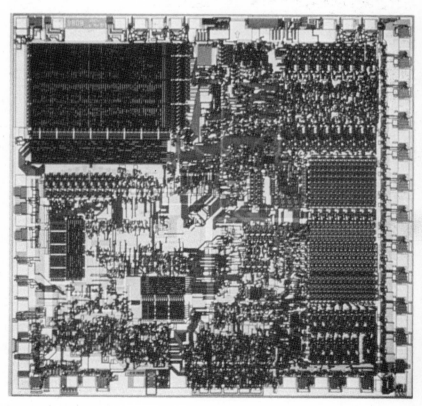

The Microcomputer

The ability to put all the functions of a central processor onto one chip (the microprocessor) has been immensely important. However, it is also possible to put a memory onto a chip. So various types of memory chip can be connected to a microprocessor, and when this is done the assemblies are called microcomputers, although they lack input and output devices.

A whole new set of abbreviations has become common parlance among electronic and computer experts. For example, a RAM is a Random Access Memory from which all information is equally accessible in a very short space of time. A RAM's contents can be modified whenever this is required. A ROM is a Read Only Memory which is programmed during manufacture and which cannot then be altered. A PROM is Programmable Read Only Memory which is programmed once after manufacture. RAMs, ROMs and PROMs are connected to the microprocessor according to whatever task the microcomputer is going to perform. A ROM coupled to a microprocessor is normally used in a pocket calculator. However, several different memories may be built into a microcomputer employed to operate the controls of a complex industrial process.

It is very important to understand that microprocessors, RAMs, ROMs and PROMs can all be standard items. They are like bricks that can be put together in different ways. Their full capacities may not always be employed but this is of little importance because they are so cheap that it is more economical to employ a standard component that will be under-used than to construct a special component.

Because of this, microcomputers can easily be built up by amateur electronic enthusiasts. In commercial and industrial uses, one microcomputer may become the core of a small business computer while several put together may control a whole manufacturing process.

But now a whole microcomputer can be created on one chip. Memories and central processor can be laid out together. Built into a standard package in the same way as a microprocessor, this miniature computer may offer us wider prospects still.

Programming

'Programming' means giving instructions to a computer before it begins work on what it must do and how the

38 A microcomputer made by Science of Cambridge, England.

work must be done. A collection of computer programs (the spelling is now universal) is called **software**. Producing software calls for considerable expertise and a great deal of time.

The microcomputer, like any other computer must be programmed, the microprocessor receiving instructions through whatever input system is employed. As we have already noted the language of the program is different from that used in the central processor or in its attendant memory or memories.

We may recall that the value of the transistor lies very much in its capacity to be used in switching systems. Many transistors are equivalent to many switches. And that is really all that a computer is. Switches are used to perform the computing activities in the ALU and switches can be used to store information in the memory. So, the number of transistors on a chip is a rough measure of its capacity in a computing system.

Computers have to work in the binary system because they are assemblies of switches. In binary any number can be described by the values 1 or 0, which can be represented by turning a switch **on** to mean 1 and **off** to mean 0.

So a program is really setting up a number of switches to go on and off to order. To do this directly would be quite impossible for all but the very simplest of operations (and even then it would be rather time consuming) so different languages have been developed which simplify programming and are automatically translated into its 'own' language (machine code based upon the binary system) by the microprocessor.

A low level programming language is one which is close to the binary system. A high level language is one which is closer to our own language. A language now in common use with microcomputers is BASIC (Beginner's All-purpose Symbolic Instruction Code). To add two **single** digit numbers together, thirty instructions would have to be produced if the operation were programmed in machine code. Only five instructions would be needed to add two **seven** digit numbers together using BASIC.

Undoubtedly programming will remain a mysterious art to many people for quite some time. But with more and more computing capacity becoming available, the possibility of programming computers in our own language is extremely attractive.

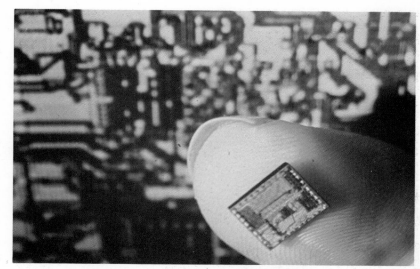

39 A microcomputer on a single chip made by Intel.

40 The main parts of Intel's microcomputer.

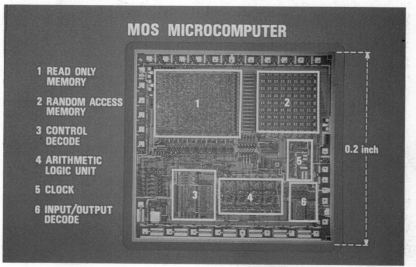

MOS MICROCOMPUTER

1 READ ONLY MEMORY
2 RANDOM ACCESS MEMORY
3 CONTROL DECODE
4 ARITHMETIC LOGIC UNIT
5 CLOCK
6 INPUT/OUTPUT DECODE

0.2 inch

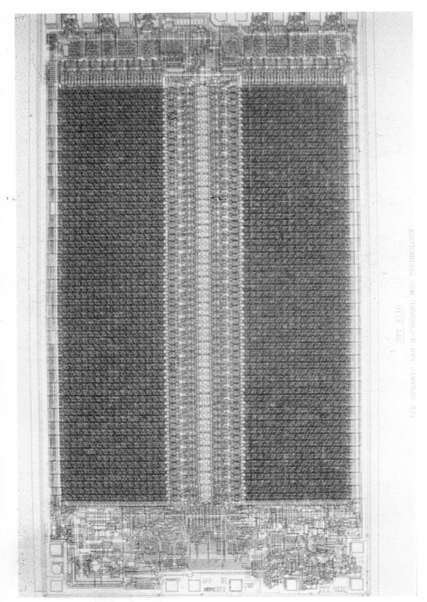

Memories

Most of us buy memories quite often! They can be books, directories, records or tapes. They are stores for information which we call upon whenever we wish. The early computers used punched paper tapes as memories. Magnetic tapes and discs followed afterwards and 'floppy discs', which are smaller than rigid discs and made in flexible plastic material, are also being used.

Whereas our own domestic tapes and cassettes are magnetised according to the volume and character of whatever has been recorded, a computer's magnetic tapes or discs carry only a series of pulses. These correspond to the binary data they have received and which they will deliver when called upon to do so by the central processor. Each pulse is equivalent to a 'bit' which, in computing language, is the smallest unit of information. The capacity of a memory can, therefore, be measured by the number of bits it will hold.

Magnetic tapes and discs can store large numbers of bits but these tapes and discs have to be moved until the information required by the central processor is brought to the reading head. Consequently, it takes some time to extract information from this type of memory.

However, memories on chips are now in use which, as well as being contained in much smaller spaces, overcome time delays. Information is stored in different places on a chip and any one of these places can be reached in the same amount of time. These memories are called location-addressable and the time required to obtain information from them – the access time – is extremely small.

Such memories employ different systems according to their purposes. For example, a ROM (Read Only Memory programmed during manufacture) employs transistors as switches that are set to correspond with the binary code. Thus a closed switch represents 1 and an open switch represents 0. A RAM (Random Access Memory) can employ transistors in a more intricate switching system or it can use transistors to store an electrical charge in small capacitors. The latter type of memory can store more information but the charge can leak away. Consequently it must be refreshed frequently. It also loses the information when switched off.

41 The circuit for a mass produced memory chip which stores 16,384 'bits' of information.

Future memories

All the advantages of packing more capacity into smaller spaces—new uses, lower costs and increased reliability—apply as much to memory chips as to special chips or microprocessors. As a result, new types of memory chip promising storage capacity of a million bits and beyond are being created.

In two kinds, the charge-couple device and the magnetic bubble memory, the bits are not located in definite positions but travel in a continuous loop, like pulses on a continuously moving and endless tape—except that there is no tape. Access time, that is the time required to pick out any series of bits, may be longer than with location-addressable systems but nothing like as long as with tapes and discs. Thus magnetic bubble memories with their potentially large capacity could very well replace tapes or discs in the near future.

Bubble memories work in an extremely intricate manner for they depend upon the fact that magnetic materials consist of many small regions which are individually magnetised, but in different directions so that normally there is no overall magnetic effect. However, these regions can be aligned in a weak magnetic field. Those which point north grow in the north-facing direction of the magnetic field. Those which point south become smaller as a result and end up like bubbles. The presence or absence of a bubble can be used to record a bit of information and the bubbles are so small that over a million can be produced on a wafer of garnet about fourteen millimetres square.

Beyond the bubble memory even more refined systems are under development offering us opportunities to store whole libraries of information in devices that we shall be able to carry about in our pockets.

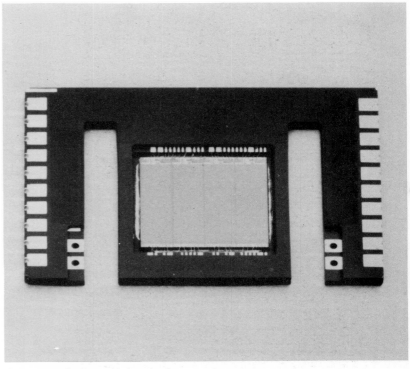

42 A garnet wafer mounted in the yoke of a magnetic bubble memory produced by Intel which can store 1,048,576 'bits' of information

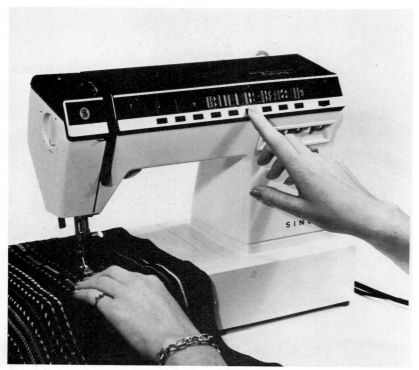

43 Singer 'Futura' sewing machine with microprocessor-based stitch selector.

Using microelectronics today

The emergence of microelectronics has been so swift that there may seem to have been little time to put it into practice. Yet there are few areas in our lives which microelectronics has not affected. A major reason for this is that microelectronic products are cheap and easily built into a wide variety of systems by economical production techniques. Another reason is that many devices employing microelectronics can be developed without spending large amounts of money and time.

The home
Whether we regard digital watches and pocket calculators as prestigious toys or necessary aids, these products are now universally available in wide variety, and we buy them avidly as we buy the many toys and games which are now appearing.

In addition, microelectronics is providing better control systems for washing, cooking and heating equipment. For some time we have had pre-set washing machines and pre-set cooker controls. But they have been bulky and expensive to make by comparison with the microelectronic systems which are now taking over. The latter are cheaper, provide more programs and are more reliable.

At present, cookers, washing machines and dishwashers are being fitted with their own microelectronic programmers. But the standard microprocessor and memories which are already available could enable us to program all the services we use in the entire house e.g. to regulate the heating system in the most efficient manner, switch on lights, draw curtains, switch on the television set for selected programmes, record them, record telephone messages and produce automatic reminders of appointments, all from one central control unit. Is all this absurdly luxurious? Barely a hundred years ago, switching on electric light would have seemed a dream to most people.

Whether we regard devices like Prestel and personal computers as 'leisure products' or not depends very much upon our personal interests. The same can be said for sewing machines, cameras and electronic organs. Like programmable heating systems or washing machines these can undertake more tasks more precisely when they are fitted with microelectronic controls. Better control of all our domestic services, more information to educate ourselves and more entertainment are, perhaps, the three main benefits which micro-electronics is now bringing into our homes.

Shopping

Whether we shop at 'the corner store' or in the local supermarket, the service we receive may well be supported by microelectronics. The local shopkeeper may be using a weighing machine or a cash register which is equipped with a microprocessor. He may also be using a small computer to keep his stock records and accounts.

The local supermarket could be overcoming one of its drawbacks by removing congestion at the check-out points and offering us a better service by presenting us with bills that not only list the prices of the things we have bought, but also indicate what these things are.

There are two systems that fulfil this function and both depend upon identifying products with bar codes which are small patterns of parallel lines of varying thickness. These define the products and their price. In one system the check-out attendant passes a wand over the bar codes on all the purchases. In another system, developed by IBM, the purchases pass over a laser beam. In both cases, all the purchases are automatically recorded, priced and totalled.

While we receive a neat bill and pass through the check-out more quickly than with previous procedures, the warehouse in the supermarket is automatically advised of sales and re-ordering of stock can be made automatically. Changes in price are easily 'keyed' into a central computer which sends the new prices to each check-out station. In addition accounts are kept automatically.

Both our shops and our supermarkets are supplied by means of vast distribution networks involving large warehouses some of which have automatic systems for collecting and despatching goods. The control of these networks and of the equipment used in warehouses is being made increasingly with the help of microelectronic systems.

Then, to go shopping at all, we may need to take some cash with us. If we get it from a bank we may well get it from a cash dispenser which employs microelectronics to check our accounts before we are served. But cash may well be going out of fashion as automatic systems for crediting and debiting our accounts become available.

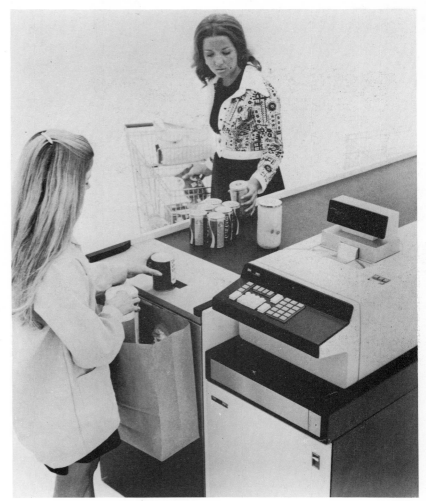

44 Products are priced and totalled automatically at this supermarket check-out point developed by IBM.

45 A telephone answering system which can be used for the remote control of office equipment and for automatic contact between offices over long distances, made by Shipton Communications Ltd.

The office

The office is an information factory. And like any factory its efficiency depends upon receiving, processing and dispensing its input – letters, accounts and records – without error and without wasting time and resources.

The computer made a considerable difference to systems employed by large offices from the nineteen-sixties onwards. Very large offices would have their own computers to handle trading accounts and wages. Smaller offices might share the use of a computer. Large computers are still used in this way.

Nevertheless, due to microelectronics, even the smallest business can now employ a computer. Farmers, garage proprietors, office managers in small companies doctors, dentists and solicitors can all choose between a variety of desk-top computers which are readily available.

Because of microelectronics, a new type of office machine has appeared – the word-processor. The word-processor relieves some typists of the boring and un-satisfying side of work, such as producing standard letters with only minor variations. The word-processor can take it over. Moreover, it can also be programmed to store records, keep accounts and help with editing. It might, therefore be better known as an office-processor.

Word-processors will be as common as filing cabinets in the not very distant future. And even the filing cabinets may go, to be replaced by microelectronic devices which will present stored information on tele-vision type screens.

One of the vital needs in any office is good com-munication. New and more easily installed telephone systems employ 'multiplexing', which is a way of routing a large number of messages over one circuit by breaking them up into pulses, a process which would not be econ-omical without the use of microelectronics. Other types of communication equipment employing microelectronics include dictating machines that collect letters from a number of managers and distribute them automatically to a team of typists, devices for activating systems such as burglar alarms by telephoning from elsewhere and even desk-top diaries which automatically provide their users with a daily presentation of commitments and appointments.

In fact, as more refined forms of communication develop, more people may be able to work at home. Indeed, if there is any worry about redundancy, it is the office rather than its staff that may suffer!

Communication

Fast push-buttoning on telephones and facilities for linking a number of callers together are now possible with the help of microelectronics. Telephone exchange equipment can be smaller and automatic testing of lines and equipment carried out more easily.

We can also get more service from our telephone if we wish. For example, we can carry a small bleeper which will let us know if we are wanted urgently. We can then go to the nearest telephone and call up the source of the bleeper's signal. Paging systems have been employed already inside hospitals and large companies but now 'radio-paging', as it is called, can be used outside them.

Some of us may be delighted to learn that pay phones are now available which give change if a call period has not been entirely used up. Others who like spending their leisure time on the telephone may be less happy to learn that they can be told how much a call is costing by means of a modestly priced unit obtainable in any multiple store. Single button operation is also available. This means that we can make the numbers between 0 and 9 into codes for very important numbers such as those belonging to our near relatives, the doctor, the police and so on – an excellent facility for emergency calls.

Broadcasting is also benefiting from microelectronic applications, in, for example, control equipment and transmitters. We too, the viewers and listeners, may benefit. The BBC has shown that programmable radios and television sets can be produced. Thus a receiver could be set, say at the beginning of a week, to pick up all the programmes we wished to hear or watch. And microelectronic-based recording systems could also store them for us.

Satellite television whereby programmes can be received not just from three or four broadcasting stations, but from thirty, forty or more is already being employed in the USA. Whether, however, we want to be drenched by world-wide television is debatable!

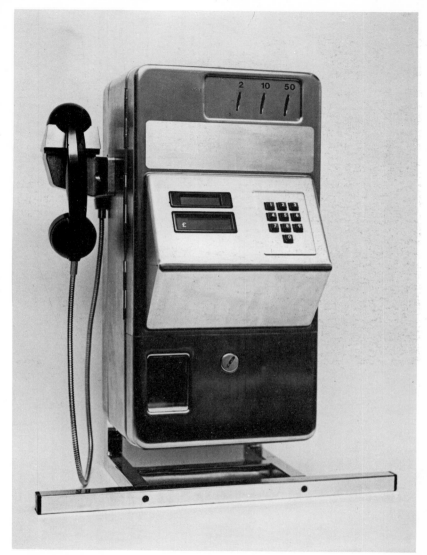

46 A coin box put into service recently by Post Office Telecommunications which gives change when a call period is not fully taken up.

47, 48 Microelectronics is making work easier in many ways. An attendance recording system introduced by Plantime Ltd helps staff and management to organise work effectively. A micrometer developed by Patscentre for Moore and Wright Ltd gives measurements in digital form.

Production

It will be a pity if the robot becomes the only symbol of microelectronic developments in manufacturing. Certainly so-called robotic techniques will be used increasingly in the large-scale manufacture of cars, television sets, washing machines and the like. But other influences upon production are equally if not more important than robots.

Electronic systems were used in automated production systems well before microelectronics appeared. Chemicals and foodstuffs were being made by automated processes from the mid-nineteen-fifties onward. What microelectronics has done is to provide much more sensitive control systems and to enable automated systems to be employed where smaller quantities are handled.

This influence is also being felt in the manufacture of metal parts that are made in small or batch quantities. Not so long ago such items could be made only by automatic techniques in very large quantities. Now batch production procedures can be tackled with the help of microelectronics. The assembly of such items in small quantities can also be accomplished more easily by what have come to be known as 'pick and place' mechanisms. These are elementary types of robot which even a small manufacturer can employ.

For full efficiency machines have to keep going. The breakdown of a comparatively minor unit, say a pump in a cooling system, can cause a whole production plant to shut down. Microelectronics enables 'condition monitoring' techniques to be employed whereby machines are provided with nerve systems which can tell how they are behaving so that likely failures can be predicted and avoided, perhaps automatically for example, by switching from one production line to a back-up line. Condition monitoring or as some engineers are calling it, health monitoring, will help further to convert the jobs of many people in industry from doing manipulative work to simply checking that it is being done satisfactorily.

The production and excellence of an enormous range of products depends upon how they are designed. Computer-aided design has been regarded by many designers as something which is only really worthwhile when designing complex products like aircraft or large structures like suspension bridges. Now, however, microelectronics enables any designer to use a computer – and not only in engineering. Patterns on fabrics and carpets, furniture shapes, the structures of buildings as well as the performance of fairly simple engineering products like pumps and valves can all be determined by the small but powerful computers which microelectronics has created. So, in design as well as in production, microelectronics is having effects which are changing traditional attitudes not just to how people work but how they are trained to do this work.

Transport

On land, sea and in the air, microelectronics is helping us to travel more comfortably, more safely and with less waste of energy. As regards motor vehicles, the more efficient use of energy is clearly of growing importance and in this respect the control of fuel and ignition systems to suit road and speed conditions is already being undertaken by microelectronic equipment. The reduction of exhaust pollution and of noise are also desirable objectives and both are being actively pursued with the help of microelectronics. Safety, a vital requirement, can be improved by more sensitive, non-skid braking systems, while radar based control for foggy conditions is a possibility. And microelectronics can build more services into our vehicles. These may help us to find our destinations or inform us of the most economical speeds at which we should travel to reach a particular destination at a particular time taking into account our position and the road conditions.

Condition or health monitoring systems can now be employed in all types of transport as well as in industry. The performance of locomotives and signalling systems can now be examined continuously so that higher levels of safety and reliability can be reached. Very soon railway passengers will pass through ticket barriers that will not delay them so much as current ticket collecting systems do in rush-hour conditions. They will also be able to buy tickets to any destination on a pay-train rather than just to stations which the train serves.

49 New shapes for aircraft to increase manoeuvrability, like this design by Rockwell International, USA, are possible with microelectronic control systems.

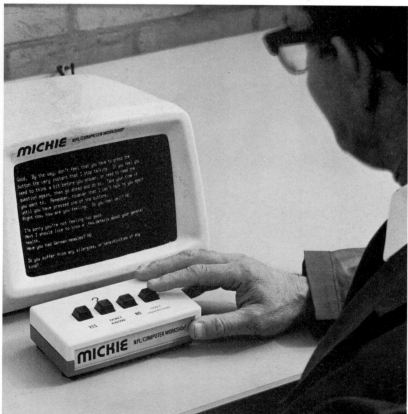

50 'Mickie' developed by the National Physical Laboratory and produced by Computer Workshop Ltd can help people to diagnose their own complaints.

Microelectronics can improve the navigational and control systems of ships as much as of aircraft for which so much microelectronic development work was first undertaken. But now control capabilities have become so great that additional control surfaces can be built on to aircraft to be operated by computer-based systems. So called 'fly-by-wire' techniques are enabling novel shapes of aircraft to be conceived and some are already being tested.

Fascinating as these developments may be, most of us will be more directly concerned with having comfortable, economical, safe and reliable transport systems. Microelectronics is helping to achieve all these objectives.

Medical equipment

If we monitor the 'health' of machines more effectively with the help of microelectronics then we can do the same for our own health. This is exactly what is happening. In hospitals, patients with heart, lung or other problems can be kept under constant surveillance by instruments which all record and interpret data as well as sound alarms if serious malfunctions occur. The local doctor's records can be regarded as forms of health monitoring. They can now be stored in a simple computer which is programmed to call attention to past illnesses to help a current diagnosis. And we can help diagnose our own complaints by answering questions put by a small computer which then provides a print-out which we can present to the doctor, thereby saving his time and starting him off on the right track.

Better instruments for testing sensory functions, like seeing and hearing, are being rapidly developed with the aid of microelectronics. One system, for example, can test the hearing abilities of newborn babies, a valuable facility since undetected deafness in babies can cause much anguish both to them and to their parents.

Highly advanced instruments, especially those which can locate tumours with greater accuracy than was possible in the past, are helping us to establish medical techniques which were quite impossible even ten years ago. They largely depend upon microelectronic systems to do their work.

Education and research

Microelectronics is affecting our education at all ages. Toys that help to teach the very young at home can be supplemented in classrooms by small computers that can present information as children want it, so preventing them from feeling anxious that the teacher may not be pleased with their work. Carefully produced and colourful programs can relieve the teacher of some of the difficulties involved in getting their pupils through prescribed curricula. They can, therefore, become more concerned with their pupils as people and not as pieces of educational blotting paper.

Various keyboard operated computers with VDU's are now available for educational purposes and one at least uses a conventional television set. Learning at home with this type of equipment will therefore become much easier and more pleasant. Then computer-based programs produced by a central educational unit, such as the Open University, will also extend domestic learning techniques.

Education and training for scientific and technological activities is already being improved owing to the availability of more powerful computers and of more sensitive instruments with which to do research work. There is hardly an instrument relying upon electronic processes for its functions which does not now employ a microelectronic system. From the humble voltmeter to the advanced electron-beam microscope, microelectronics is making instruments more precise and, in many cases, more versatile. Many such instruments are, of course, needed in the study of microelectronics itself and in its further development.

51 Learning about microelectronics with PET, a personal computer made by Commodore Business Machines Ltd.

Two success stories

Considerable enterprise has been shown by manufacturers in taking up microelectronics so quickly and
this enterprise has not been confined to those making
complex and so-called 'high technology' products. As
the first of the following case histories shows microelectronics can be employed to great effect in comparatively simple devices.

A microelectronic taximeter

About three years ago, three London taxi proprietors,
Ronnie Samuels, Ronnie Richmond and Raymond Allan,
were disturbed because taximeters could be obtained
only from two sources – both foreign. They were also
tired of having to change the tariffs on all their taxis at
frequent intervals since the high level of inflation was
causing a rapid rise in costs. A mechanical taximeter is
accurate and reliable but changing its tariffs is not an
easy job.

Could an all-electronic meter be made? A first thought
was to produce a digital electronic counter showing
units, not fares, and to have simply a chart giving prices
at current tariffs for the number of units shown. However, the Public Carriage Office insisted that a meter
must show the actual cost of any journey.

The three friends therefore drew up a specification
for an ideal taximeter, formed a company, Cavalier
Engineering, and teamed up with Randall Electronics,
a company specialising in time switches for heating
systems. This association led them to create their ideal
meter, one based upon a standard microprocessor
made by Texas Instruments. The microprocessor re-
ceives details of distance travelled and journey time
from the conventional drive and a clock. It then works
out the cost of the journey having received the tariff
from a simple plug-in printed circuit board which can be
easily changed. The cab operator can now tell how many
journeys a cab has made, how much it has earned and
how many miles it has travelled while the customer gets
a clear indication of how much he has to pay.

Several other microprocessor-based taximeters are
now available. Nevertheless, the enterprise of three
men, their awareness of a need through familiarity with
a particular activity, and their willingness to use the
services of experts enabled them to create an extremely
profitable venture.

52 A microelectronic taximeter–introduced by
three taxi service operators who were
dissatisfied with the traditional meter.

A pocket thermometer

Of course, many manufacturers will wonder how much it costs to develop a microelectronic product. There can be no general guide. Nevertheless, the following story may interest them. It should also tell all of us a little about how products, upon which some of us depend for our livelihood, are created.

Comark Electronics Limited is a small company making instruments. By 1973 it was well known for its dial-reading thermometers but a new management team wanted new products to expand the company Microelectronics might help. For example, would it be possible to make a pocket digital thermometer with an integrated circuit? Comark's engineers thought it would.

A company in the GEC group, Britain's largest manufacturer of electrical and electronic products, agreed to develop and make a special integrated circuit for £15,000 and the rest of the design and development work was undertaken by two Comark engineers. They exceeded their original estimates of development time and cost but the final figures were three years for the time taken and £60,000 for all development costs, including those associated with preparing for manufacture and setting up testing systems for the production versions.

No serious difficulties were encountered either with developing the thermometer or with setting up the production systems. These were arranged to employ automatic assembly and testing procedures. Consequently, production proceeded without any great increase in staff. Comark also established detailed inspection procedures and prepared a comprehensive technical service manual.

The thermometer recovered its development costs in three years. In addition it helped to multiply the company's turnover eighteen times and to encourage management to make even more of microelectronics.

The two stories suggest that there is much to be made from microelectronics in many sectors of industry. The key need is enterprise.

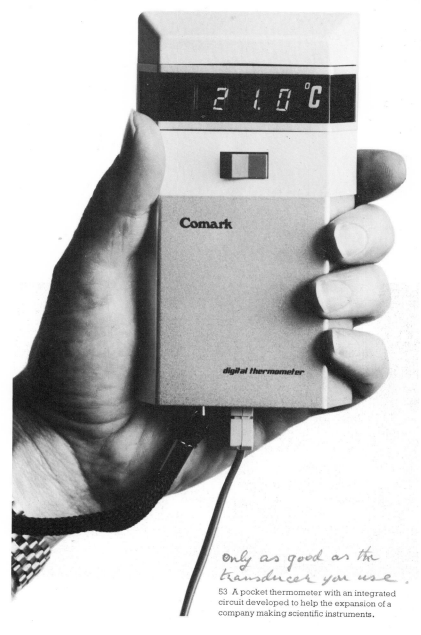

only as good as the transducer you use.

53 A pocket thermometer with an integrated circuit developed to help the expansion of a company making scientific instruments.

The way ahead

Forecasting the future is a risky business. Technical prediction is one thing : actual use is another. Progress depends upon attitudes and economics as well as upon science and technology. In microelectronics, as elsewhere, we shall adopt what we can use and afford.

Microelectronics experts all agree that it will be possible to put more components on a chip. We are ready to put more and more computing capacity into smaller and smaller spaces and, as a result, to handle more and more information. One expert has said that by 1985 we could tap a vast range of information sources, do all our calculations, and get in touch with our friends anywhere in the world, using a device in our pocket.

As we put more components on a chip our production techniques may change. For example, the wavelength of light limits the size of the apertures or 'windows' in the photolithographic process. Yet there appears to be no difficulty in using X-rays which have a shorter wavelength than light.

There is probably no need to make chips themselves much smaller. However, even more refined ways of making them are bound to be developed. The microelectronics industry will be more and more capital-intensive which means that the cost of manufacturing equipment will become even greater than the cost of the labour of those employed to operate it. Increasing costs rather than technical considerations might, in consequence, slow down the race to pile more and more capacity onto the chip.

Even so, microelectronic developments will help us to handle more and more information. As optic fibre cables take over from our conventional telephone lines we shall be able to transmit all this extra information without difficulty, using it to communicate more readily, to control our machines more precisely, to educate ourselves more easily, and to measure natural phenomena like the weather and soil conditions more accurately so as to diminish the hazards to life.

What are our priorities?

We are very conscious at the present time of our need to conserve energy, especially energy from fossil fuels, which will not last for ever.

Microelectronics cannot give vehicles, aircraft and machines greater efficiency than that determined by the principles on which they work. But many of these products consume more energy than they need for a substantial amount of their working lives. More precise tailoring of a machine's performance to the work it is doing can certainly be aided by microelectronics. It is beginning to happen in cars and should happen increasingly in future.

Tailoring supply to demand according to the operating conditions certainly applies to keeping ourselves warm and comfortable. Microelectronics is already helping here.

As we grapple with our energy needs we must also take account of our environmental problems. Many of the answers to these problems depend upon continuous measurement and analysis of pollutants in particular. When agreed danger levels are reached corrective action can then be taken automatically. Since microelectronic systems for coping with this type of activity can be robust and inexpensive, we should expect to see a considerable network of environmental control devices set up within the next decade.

The detection of undesirable conditions is of course the prime purpose of medical diagnosis. Microelectronics will certainly help us increasingly in this respect. And it should, we hope, enable more efficient medical treatments to be employed in the poorer countries of the world.

Finally, we must use microelectronics to help us learn more about ourselves and the world about us. Like other great technological developments, such as rail travel and broadcasting, it will change our ways of living and, indeed, our ways of thinking. Knowing more about ourselves and our world, we shall be able to make the changes we want – and that is a very important reason for finding out more about microelectronics and what it can do.

REFERENCES

Microelectronics, Robert N. Noyce.
Microelectronic Circuit Elements, James D. Meindl.
The Large-Scale Integration of Microelectronic Circuits,
William G. Oldham.
Microelectronic Memories, David A. Hodges.
Microprocessors, Hoo-Min D. Toong,
all from 'Scientific American' September 1977 Vol. 237
Number 3, published by Scientific American Inc.

Revolution in Miniature, Ernest Braun and Stuart
MacDonald. Cambridge University Press 1978.

The Transistor, Joachim Dosse. D. Van Nostrand
Company Inc 1964.

The Collapse of Work, Clive Jenkins and Barrie Sherman.
Eyre Methuen Ltd 1979

The Future with Microelectronics, Ian Barron and Ray
Curnow. Frances Pinter (Publishers) Ltd 1979.

ACKNOWLEDGEMENTS

Considerable assistance was received in preparing
and editing this book from Dr Frank Greenaway and
Mr V. K. Chew of the Science Museum and also from
Ms Jane Raimes.

Valuable advice was given by Mr Stan Mash at
Mackintosh Publications Ltd and contributions were
received from Mr Howard Kornstein of the Intel
Corporation, Mr Roger Whitehead of Mullard Ltd,
Mr Richard Waller of CAP Ltd and Mr Gerald Lane of
the Design Council.

We would like to thank PPR International Ltd and
PA Management Consultants Ltd respectively for data
on the Cavalier and Comark projects.

PICTURE CREDITS

We would like to thank the following, who have kindly
provided illustrations :
A R C Ltd (47)
Barnaby's Picture Library (16, 19, page 35)
Casio Electronics Co Ltd (12)
Computer Games Ltd (13)
Department of Industry (4, 9, 25, 27, 31, 32, 33, 34)
Design Council (8, 52, 53)
I B M United Kingdom Ltd (44)
Intel Corporation (37, 39, 40, 41, 42)
Microelectronics Journal, Sept. 1978 (7)
Moore & Wright Ltd (48)
Mullard Ltd (14, 21, 24, 28, 29, 30, 35)
National Coal Board (18)
National Physical Laboratory (50, Crown Copyright)
Post Office Telecommunications (11, 46)
Rockwell International, via Flight Magazine (49)
Science Museum (2, 3, 5, 6, 10, 38, 51, page 34)
Shipton Communications Ltd (45)
Singer Co (UK) Ltd (43)
Standard Telephones & Cables Ltd (20)
Unimation (Europe) Ltd (17)

Printed in England for Her Majesty's Stationery Office by
Raithby, Lawrence & Company Limited at the
De Montfort Press : Leicester and London
Dd 596343 K160